BACKYARD MEAT RABBITS

From Choosing Breeds To Sustainable Practices: A Comprehensive Guide To Raising, Daily Care, Feeding, Nutrition, Harvesting, And Sustaining A Healthy Rabbitry

ETHAN HARRY

Table of Contents

CHAPTER ONE 5

INTRODUCTION TO BACKYARD MEAT RABBITS 5

Understanding The Appeal Of Meat Rabbits 5

History Of Rabbit Farming 9

Benefits Of Raising Meat Rabbits............................13

Overview Of The Guide..16

CHAPTER TWO 21

CHOOSING THE RIGHT BREEDS 21

Common Meat Rabbit Breeds.................................21

Characteristics Of Meat Rabbit Breeds..................24

Selecting Breeds Based On Climate And Conditions..............28

Where To Purchase Your Rabbits 33

CHAPTER THREE 38

SETTING UP YOUR RABBITRY 38

Selecting A Suitable Location..................................38

Building Or Buying Rabbit Hutches 40

Essential Equipment And Supplies.........................43

Ensuring Safety And Protection From Predators..................47

CHAPTER FOUR 52

RABBIT CARE AND MANAGEMENT 52

Daily Care Routine 52

Feeding And Nutrition 55

Health And Disease Management 58

Handling And Socializing Your Rabbits 62

CHAPTER FIVE 65

BREEDING MEAT RABBITS 65

Understanding Rabbit Reproduction 65

Selecting Breeding Stock 68

Mating And Gestation 72

Caring For Pregnant Does And Newborn Kits 76

CHAPTER SIX 80

GROWING AND HARVESTING 80

Growth Stages Of Meat Rabbits 80

Optimal Conditions For Growth 84

Determining The Right Time For Harvest 87

Humane Methods Of Harvesting 90

CHAPTER SEVEN 95

PROCESSING RABBIT MEAT 95

Preparing For Processing 95

Step-By-Step Processing Guide 99

Meat Cuts And Preparation 102

Storing And Preserving Rabbit Meat 107

CHAPTER EIGHT 111

SUSTAINABLE AND ETHICAL PRACTICES 111

Ethical Considerations In Rabbit Farming 111

Implementing Sustainable Practices 115

Utilizing Rabbit By-Products 119

Engaging With The Community And Educating Others 122

THE END 126

CHAPTER ONE

INTRODUCTION TO BACKYARD MEAT RABBITS

Understanding The Appeal Of Meat Rabbits

Raising meat rabbits in your backyard has become increasingly popular due to several compelling reasons. Firstly, rabbits are remarkably efficient at converting feed into meat. Their rapid growth rate means they can be ready for harvest within a mere 10 to 12 weeks. Additionally, rabbits have an impressive reproductive rate, with a single doe (female rabbit) capable of producing multiple litters annually. Each litter typically consists of several kits (baby rabbits), allowing a small backyard operation to yield a substantial amount of meat throughout the year.

One of the major attractions of raising rabbits is their relatively low cost and minimal space requirements. Unlike larger livestock, rabbits can be raised in small areas, making them suitable for urban or suburban environments. This is a significant advantage for those who wish to engage in small-scale meat production but lack extensive land resources. Moreover, rabbits have a relatively low environmental impact. They produce less waste and require significantly less water compared to traditional meat sources such as cattle or pigs. This makes rabbit farming an environmentally friendly option for sustainable meat production.

Rabbit meat itself is highly nutritious and lean, often considered a delicacy in various cuisines around the world. It is rich in protein, low in fat, and contains essential vitamins and minerals, making it a healthy alternative to other meats. For health-conscious individuals or those seeking to diversify their diet with a sustainable protein source, rabbit meat presents an excellent choice.

In terms of backyard farming, rabbits are relatively easy to care for. They require simple housing structures, which can be built with basic materials. Proper ventilation, protection from extreme weather, and regular cleaning are essential to maintaining a healthy rabbitry. Rabbits also have a diverse diet, primarily consisting of hay, fresh vegetables, and specially formulated pellets, which are readily available and affordable. Their feed-to-meat conversion efficiency means that the investment in their diet yields high returns in terms of meat production.

Additionally, the manure produced by rabbits is a valuable byproduct. Rabbit manure is considered one of the best natural fertilizers for gardens, as it is rich in nitrogen and other nutrients. It can be directly applied to plants without composting, unlike the manure from many other animals. This dual benefit of meat and fertilizer production makes raising rabbits an even more appealing option for backyard farmers.

History Of Rabbit Farming

Rabbit farming boasts a rich history that spans various cultures and historical periods. The domestication of rabbits can be traced back to the Iberian Peninsula, which is present-day Spain and Portugal. It was here that the Romans first recognized the potential of rabbits for domestication. The Romans appreciated rabbits for their rapid breeding capabilities and their meat, which quickly became a staple in the Roman diet. This early domestication marked the beginning of rabbit farming as a recognized agricultural practice.

During the Middle Ages, the practice of keeping rabbits evolved further. Rabbits were kept in specially designed enclosures known as "warrens." These warrens were primarily found in monasteries and the estates of the nobility. Monks, in particular, found rabbits to be valuable as a food source. Rabbit meat was deemed acceptable to eat during Lent when other meats were prohibited by the church. This religious sanction helped to popularize rabbit farming within monastic communities and noble estates across Europe. Over time, the practice of rabbit farming spread throughout the continent, becoming an integral part of medieval agriculture.

The 19th and 20th centuries saw rabbit farming gaining significant traction in North America. This period was marked by the two World Wars, during which rabbit farming was heavily promoted as a valuable food source. Governments encouraged citizens to raise rabbits in their backyards to supplement their food supply during times of rationing. Rabbits were ideal for this purpose due to their efficient meat production and low maintenance requirements. The promotion of rabbit farming during these times was part of a broader effort to ensure food security and self-sufficiency among the populace.

In contemporary times, rabbit farming continues to be practiced around the world. Both small-scale backyard operations and larger commercial farms contribute to the global supply of rabbit meat. Advances in breeding techniques, nutrition, and husbandry have made rabbit farming more efficient and productive. Today, rabbit meat is valued not only for its taste but also for its nutritional benefits, being a lean source of protein. Rabbit farming remains a versatile and sustainable agricultural practice that adapts to the needs of different regions and communities.

The history of rabbit farming is a testament to the adaptability and resilience of this agricultural practice. From its early days in the Iberian Peninsula to its widespread adoption during the World Wars, and its continued relevance today, rabbit farming has proven to be an enduring and valuable component of global agriculture.

Benefits Of Raising Meat Rabbits
Raising meat rabbits offers numerous benefits, making it an appealing option for hobbyists and serious farmers alike. One of the primary advantages is their high reproduction rate. Rabbits breed quickly and can produce several litters each year, ensuring a steady and reliable supply of meat. This prolific breeding capability allows farmers to scale up their operations swiftly and efficiently.

Another significant benefit is the efficient feed conversion of rabbits. They convert feed into meat more effectively than many other livestock animals. This efficiency translates into lower feed costs, making rabbit farming a cost-effective venture. Additionally, rabbits require relatively little space compared to larger livestock. This space efficiency makes them suitable for small backyards or even urban settings, where space is at a premium. Small-scale farmers and urban dwellers can easily manage a rabbitry without needing extensive land or resources.

Raising rabbits also has a low environmental impact. Rabbits produce less waste compared to larger animals, and their lower water usage further minimizes their ecological footprint.

This makes rabbit farming a more sustainable and environmentally friendly option. For those concerned about the environmental impact of their food production, raising rabbits can be a conscientious choice.

Nutritionally, rabbit meat is an excellent addition to the diet. It is lean, high in protein, and contains essential vitamins and minerals. Rabbit meat is also lower in fat and cholesterol compared to other meats like beef and pork, making it a healthier option for those looking to maintain a balanced diet. The nutritional benefits of rabbit meat can appeal to health-conscious consumers and those looking to diversify their protein sources.

Economically, raising rabbits presents significant opportunities. Small-scale farmers can generate income through the sale of rabbit meat, breeding stock, and by-products like fur. The demand for rabbit meat and other rabbit-related products can create a profitable niche market. Additionally, the relatively low startup costs and operational expenses make rabbit farming an accessible venture for those looking to enter the agricultural industry.

Overview Of The Guide

This comprehensive guide is your go-to resource for successfully raising meat rabbits. Whether you're a novice or an experienced farmer, this guide covers everything you need to know—from choosing the right breeds to processing the rabbits for meat. Each chapter dives deep into specific aspects of rabbit care and management, offering practical tips rooted in both traditional practices and modern techniques.

We'll start by exploring various rabbit breeds ideal for meat production, helping you select the ones that best meet your needs. The next chapters will guide you through the essentials of rabbit housing. You'll learn how to build or purchase the perfect hutches and maintain a clean, healthy environment for your rabbits. Proper housing is crucial for their well-being and productivity, and we'll provide detailed instructions on setup and maintenance.

Feeding and nutrition are vital components of rabbit care. This guide offers detailed guidance on what to feed your rabbits, including the types of feed, appropriate portions, and feeding schedules. We'll cover the nutritional needs of different life stages to ensure your rabbits grow healthy and strong.

Breeding rabbits can be complex, but we'll simplify the process for you. You'll learn step-by-step how to manage breeding, from selecting mating pairs to caring for the kits until they are weaned. This section will also address common challenges and how to overcome them, ensuring a smooth breeding process.

Health care is another critical aspect covered in this guide. We'll identify common health issues and provide strategies for prevention and treatment. By understanding the signs of illness and knowing how to respond, you can keep your rabbits healthy and productive.

Finally, the book will guide you through the harvesting and processing of your rabbits. We'll detail the humane and efficient methods for slaughtering, dressing, and preparing the meat, ensuring you get the best results from your efforts.

By the end of this guide, you will have the knowledge and confidence needed to raise meat rabbits successfully, whether your goal is to provide food for your family or to start a small business. This guide aims to equip you with the skills to manage every aspect of rabbit farming, ensuring a rewarding and sustainable venture.

CHAPTER TWO

CHOOSING THE RIGHT BREEDS

Common Meat Rabbit Breeds

Selecting the right breed is crucial when raising rabbits for meat, as different breeds offer varying benefits in terms of size, growth rate, and meat-to-bone ratio. Here are some common meat rabbit breeds that are popular among breeders:

1. New Zealand White: One of the most renowned meat rabbit breeds, New Zealand Whites are celebrated for their rapid growth and high-quality meat. These rabbits are large, typically weighing between 9 to 12 pounds. Their robust size and efficient feed conversion rate make them a top choice for meat

production. New Zealand Whites also have a calm temperament, which makes them easier to handle and raise.

2. Californian: Another excellent breed for meat production, Californians are slightly smaller than New Zealand Whites but still offer a good growth rate and impressive meat yield. They generally weigh around 8 to 10 pounds. Californians are known for their white bodies with distinct dark markings on their ears, nose, feet, and tail. Their rapid growth and efficient meat-to-bone ratio make them a popular choice among meat rabbit breeders.

3. Flemish Giant: While the Flemish Giant is primarily known as a show breed due to its massive size, it can also

be raised for meat. These rabbits are one of the largest breeds, often exceeding 14 pounds. However, they grow slower compared to other meat breeds and require more space and feed. Despite these drawbacks, their substantial size can result in a large quantity of meat, making them a viable option for some breeders.

4. Silver Fox: Known for its unique fur and excellent meat production, the Silver Fox is a medium to large breed, weighing between 9 to 12 pounds. They have a striking appearance with their dense, silver-tipped fur and a gentle, docile temperament, which makes them easy to manage. Silver Fox rabbits are valued for their good meat quality and

efficient growth rate, making them a solid choice for meat production.

5. Champagne d'Argent: An old French breed, the Champagne d'Argent is notable for its distinctive silver coat and flavorful meat. These rabbits are medium-sized, typically weighing around 9 to 12 pounds. They are known for their excellent meat quality and unique appearance. The Champagne d'Argent is a reliable breed for meat production, combining good growth rates with a desirable meat-to-bone ratio.

Characteristics Of Meat Rabbit Breeds

When selecting a meat rabbit breed, it's crucial to evaluate specific characteristics that contribute to efficient and profitable meat production. The following key traits are essential to consider:

Growth Rate: The growth rate of a rabbit breed significantly impacts the overall efficiency of meat production. Breeds that grow rapidly reach market weight more quickly, reducing the duration and cost of feeding and housing. Faster-growing rabbits allow for a quicker turnover, enabling farmers to raise more batches within the same timeframe. This trait is particularly beneficial for those looking to maximize production and profit margins.

Size and Weight: Larger rabbit breeds generally provide more meat, making them attractive for meat production. However, it's essential to balance size with growth rate. While a larger rabbit will yield more meat, it may also take longer to reach its full size. Therefore, choosing a breed that offers a good balance between size and growth rate is crucial. This ensures that the rabbits grow efficiently while providing a substantial amount of meat.

Meat-to-Bone Ratio: The meat-to-bone ratio is a critical factor in meat rabbit breeds. A higher meat-to-bone ratio means that a larger proportion of the rabbit's body weight is composed of edible meat rather than bones. This trait enhances the overall efficiency of meat production, as more of the rabbit can be utilized for consumption. Breeds with a favorable meat-to-bone ratio are preferred because they maximize the yield of edible meat from each rabbit.

Temperament: The temperament of the rabbit breed is another important consideration, particularly for those new to rabbit farming. Calm and easy-to-handle rabbits are much easier to manage on a daily basis. A good temperament makes routine care, feeding, and handling less stressful for both the rabbits and the farmer. Breeds known for their docile nature are ideal for creating a manageable and pleasant farming environment.

Health and Hardiness: Robust health and hardiness are vital traits for any meat rabbit breed. Breeds that are resistant to common diseases and capable of thriving in various environmental conditions reduce the risks associated with rabbit farming. Healthy, hardy rabbits require less medical intervention and maintenance, which lowers overall costs and improves productivity. Selecting breeds known for their strong health and adaptability ensures a more sustainable and successful meat production operation.

Selecting Breeds Based On Climate And Conditions

When selecting rabbit breeds for your environment, it's important to consider how well they adapt to your local climate and living conditions. Here's a guide to help you choose the right breed:

Hot Climates

In regions with high temperatures, it's crucial to select rabbit breeds that can tolerate the heat. Rabbits are generally susceptible to heat stress, so opting for breeds with lighter coats is beneficial. For example, the New Zealand White is a popular choice for hotter climates due to its light fur, which helps it stay cool. Additionally, providing adequate shade and ventilation is essential to prevent overheating.

Cold Climates

Conversely, if you live in an area with harsh winters, you'll need breeds that can withstand the cold. Rabbits with thicker fur coats are better suited for these conditions as they provide natural insulation against low temperatures. Breeds such as the Californian and Flemish Giant have dense fur that helps them stay warm during winter months. Ensuring that their housing is well-insulated and protected from the elements will further enhance their comfort and health.

Housing Conditions

Your available space and housing setup are also significant factors in breed selection. Some rabbit breeds require more space due to their larger size. For example, giant breeds like the Flemish Giant need more room to move around comfortably. Assess your current or planned housing to ensure it can accommodate the breed's space requirements. Adequate housing is crucial for the rabbits' well-being and can impact their growth and productivity.

Local Availability

Sometimes, the most practical breed to choose is the one that is locally available. Rabbits bred and raised in your area are likely already adapted to the local climate and conditions, making them easier to care for. Local breeders can provide valuable insights into the best practices for raising these rabbits in your specific environment. Moreover, purchasing locally can support the community and reduce transportation stress for the animals.

Purpose

While your primary goal might be meat production, it's worth considering breeds that offer additional benefits. Some breeds are known for their dual purposes, such as producing high-quality fur or being suitable for show purposes. For instance, the Rex breed is renowned for its luxurious fur and can be raised for both meat and fur production. Similarly, the American Chinchilla is valued for its meat and show qualities. By choosing a breed that meets multiple needs, you can maximize the benefits and versatility of your rabbitry.

Where To Purchase Your Rabbits

Purchasing your meat rabbits from a reputable source is crucial to the success of your rabbit farming venture. Here are some comprehensive tips on where and how to buy your rabbits to ensure you start off on the right foot:

Local Breeders: Buying rabbits from local breeders is often the best choice. It allows you to visit their facilities and inspect the conditions in which the rabbits are raised. By doing so, you can choose healthy, well-cared-for animals. Additionally, local breeders are usually willing to offer valuable advice and ongoing support, which can be incredibly beneficial as you start your rabbit farming journey. Establishing a relationship with a local breeder can provide you with a reliable source of rabbits and valuable insights into rabbit care and management.

Rabbit Shows and Expos: Attending rabbit shows and expos is another excellent way to purchase your rabbits. These events bring together numerous breeders and showcase a variety of rabbit breeds. You can meet breeders face-to-face, see their animals, and often buy rabbits directly at these events. Purchasing rabbits at shows ensures that you are getting high-quality animals, as breeders often display their best stock. Moreover, these events provide an opportunity to learn more about different breeds and rabbit farming practices.

Online Breeders and Farms: The internet has made it easier to find and purchase rabbits from breeders and farms across the country. However, buying rabbits online requires thorough research to ensure you are dealing with a reputable seller. Look for online reviews, ask for references, and check if the breeder provides proper health guarantees for their rabbits. While purchasing online can be convenient, it is essential to verify the breeder's reputation to avoid potential issues with the health and quality of the rabbits.

Agricultural Extension Services: Local agricultural extension services can be an excellent resource for finding reputable rabbit breeders. These services often have connections with local breeders and can provide recommendations based on your specific needs. Additionally, extension services can offer resources and support for starting and managing your rabbit farm, making them a valuable partner in your farming journey.

Rabbit Associations: National and regional rabbit associations are reliable sources for finding reputable breeders. These associations typically maintain directories of breeders who adhere to high standards of care and breeding practices. By consulting these directories, you can find breeders who are committed to producing healthy, well-bred rabbits. Associations can also offer additional resources, such as training, networking opportunities, and industry updates, which can be beneficial for your rabbit farming operations.

CHAPTER THREE

SETTING UP YOUR RABBITRY

Selecting A Suitable Location

Picking the right place for your rabbitry is important for keeping your rabbits healthy and making your work easier. Here are some things to think about:

Space

Make sure you have enough room for all the rabbits you want to keep. You'll need space for their cages, food, and supplies.

Accessibility

Choose a spot that's easy to get to every day for feeding, cleaning, and taking care of the rabbits. If you plan to have a lot of rabbits, think about how trucks or larger equipment will access the area.

Environment

Rabbits don't like very hot or very cold weather. They need a place that has good airflow and shade. Don't pick a spot that's too windy or always in direct sunlight. A place with some natural shade, like under trees, is a good choice.

Drainage

It's important to have good drainage to keep water from building up and making the area muddy. Muddy conditions can make rabbits sick. The ground should be a little higher or have a way to drain water properly.

Quiet

Rabbits get scared by loud noises. Find a quiet place away from busy roads, loud machines, or other animals that could scare them.

Building Or Buying Rabbit Hutches

Rabbit hutches are where your rabbits will live. You can either make your own or buy one that's already made. Here's what you need to think about:

Size and Design:

Each rabbit needs enough space to move around easily. A good rule is to have a hutch that lets the rabbit take at least three hops in any direction. Multi-level hutches can give more space without taking up too much room.

Materials:

Use strong, safe materials. Wood is common, but make sure it's treated so it can withstand the weather. Wire mesh is important for air to flow through, but the gaps should be small so the rabbits' feet don't get hurt.

Protection:

The hutch should have solid walls and a roof to keep out wind, rain, and predators. Make sure there's a part of the hutch that's closed off and insulated to keep the rabbits warm.

Cleanliness:

The hutch should be easy to clean. Having removable trays under a wire floor can catch waste, making it easier to keep the hutch clean and sanitary.

Choosing The Right Hutch

Whether you decide to build your own hutch or buy one, consider these important factors to ensure your rabbits have a safe, comfortable, and clean living space.

Building Your Own Hutch:

Building your own hutch can be a rewarding project and allows you to customize the size and design to fit your needs. Make sure to use weather-resistant wood and safe wire mesh with small gaps. Plan for a section of the hutch that is fully enclosed and insulated to protect your rabbits from the elements and predators. Also, design the hutch so it is easy to clean, with features like removable trays for waste.

Buying a Pre-made Hutch:

If you decide to buy a hutch, look for one that meets all the criteria for size, materials, protection, and cleanliness. Pre-made hutches can save you time and effort, and many are designed with features that make them easy to maintain. Ensure the hutch is spacious enough for your rabbits to move around comfortably and has a section that provides shelter from the weather and predators.

Essential Equipment And Supplies

To maintain a smooth-running rabbitry, having the right equipment and supplies is essential. Here's a detailed list of what you'll need:

Feeding and Watering

Ensuring your rabbits have a constant supply of fresh food and water is crucial for their health and well-being. Use sturdy, tip-proof bowls or invest in automatic feeders and waterers. These help prevent spills and contamination, ensuring that your rabbits always have access to their essentials. Automatic systems can be particularly useful in larger rabbitries as they save time and reduce the frequency of refilling.

Bedding

Providing clean, absorbent bedding is important for the comfort and hygiene of your rabbits. Options include straw or wood shavings. However, avoid using cedar shavings, as the aromatic oils can be harmful to rabbits, potentially causing respiratory issues or skin irritations. Regularly change the bedding to maintain a clean environment and minimize the risk of infections and parasites.

Nesting Boxes

If you plan on breeding rabbits, nesting boxes are an absolute necessity. These boxes offer pregnant does a secure, comfortable place to give birth and care for their kits. Nesting boxes should be placed in the rabbit's cage a few days before the doe is due to give birth. Ensure they are clean, dry, and filled with suitable nesting materials like straw or hay. After the kits are born, regularly check and clean the boxes to maintain a healthy environment.

Health Supplies

A basic first-aid kit is an important part of any rabbitry. Include antiseptic, nail clippers, and any necessary medications. Regularly check your rabbits for signs of illness or injury, such as changes in appetite, behavior, or appearance. Early detection and treatment can prevent minor issues from becoming major health problems. Consider consulting with a veterinarian who has experience with rabbits to ensure you have the correct supplies and knowledge to care for your animals properly.

Grooming Tools

Grooming is essential for maintaining the health and appearance of your rabbits, especially for breeds with long or dense fur. Depending on the breed, you may need various grooming tools such as brushes, combs, and nail clippers. Regular grooming helps prevent matting, reduces shedding, and allows you to check for parasites or skin conditions. Additionally, it can be a good bonding experience for you and your rabbits, making them more comfortable with handling.

Ensuring Safety And Protection From Predators

Protecting your rabbits from predators is crucial when you're running a rabbitry. Here are some important steps to keep them safe:

Securing Hutches

Firstly, it's essential to ensure that all hutches are safe from predators. Use strong locks and secure latches that predators like raccoons and foxes can't easily break. Check regularly for any weak spots that could be vulnerable to attacks.

Using Fencing

If your rabbits have an outdoor area to roam, enclose it with sturdy fencing. This fencing should be tall enough to prevent animals from jumping over it and extend underground to stop burrowing predators from getting in. This extra depth makes it harder for animals like foxes to dig their way inside.

Supervising Playtime

Always supervise your rabbits when they're outside of their hutches. Even if the area seems secure, it's safest to keep an eye on them. This way, you can quickly respond if any predators are spotted nearby or if there's any other danger.

Adding Lights or Alarms

Consider using motion-activated lights or alarms around your rabbitry. These can scare away predators that might come out at night. Place them strategically along the perimeter of your rabbitry to provide an extra layer of protection when it's dark.

Regular Checks

Regularly inspect your hutches, fencing, and the surrounding area for any signs of damage or potential entry points for predators. Fix any issues promptly to maintain a secure environment for your rabbits.

Safe Handling

When handling your rabbits, be cautious and gentle. Avoid sudden movements or loud noises that could startle them, as this might make them more vulnerable to stress or accidents.

Emergency Plans

Have a plan in place for emergencies, such as what to do if a predator is spotted or if there's a breach in security. Being prepared can make a big difference in protecting your rabbits.

Consulting Experts

If you're unsure about the best ways to protect your rabbits from predators, consider seeking advice from experienced rabbit owners or animal experts. They can offer valuable insights and tips based on their own experiences.

☐

CHAPTER FOUR

RABBIT CARE AND MANAGEMENT

Daily Care Routine

Maintaining a daily care routine for rabbits is crucial for their health and well-being. This routine encompasses several essential tasks that ensure rabbits remain healthy, happy, and comfortable.

First and foremost, providing fresh water daily is vital as rabbits require constant access to clean hydration. This involves checking and refilling water bottles or bowls to ensure they always have enough to drink.

Another integral part of daily care is maintaining a clean living environment. This includes regular cleaning of their hutch or cage. Removing soiled bedding, such as hay or wood shavings, and replacing it with fresh material helps to prevent odor buildup and maintain a hygienic space for the rabbits. A clean environment also reduces the risk of health issues arising from bacterial growth or contamination.

In addition to cleanliness, monitoring and replenishing their food supply is essential. Rabbits should have a consistent diet of hay, pellets, and fresh vegetables suitable for their age and health requirements. Daily inspection of their food ensures they are receiving adequate nutrition and helps prevent issues related to diet imbalance.

Grooming plays a significant role in a rabbit's daily care routine, especially for breeds with longer fur. Regular brushing with a gentle comb or brush helps to remove loose fur, prevent matting, and maintain their coat's health and appearance. It also provides an opportunity to bond with the rabbit and check for any skin abnormalities or parasites that may require attention.

Checking and trimming their nails as needed is another critical aspect of rabbit care. Overgrown nails can cause discomfort and potentially lead to health problems, so keeping them trimmed to a suitable length is important for their well-being.

Beyond physical care, spending quality time with rabbits daily is beneficial for their socialization and mental stimulation. Interacting with them helps build trust and allows caregivers to observe any changes in behavior or health that may indicate underlying issues needing veterinary attention.

Feeding And Nutrition

Feeding rabbits a balanced diet is essential to their overall health and well-being. The foundation of a rabbit's diet should primarily consist of hay, which serves several critical purposes. High-quality hay not only aids in proper digestion but also helps maintain dental health by naturally wearing down their teeth as they chew. This continual chewing action prevents dental issues that can arise from overgrown teeth.

In addition to hay, fresh vegetables play a crucial role in providing essential vitamins and minerals to rabbits. Leafy greens such as kale, spinach, and romaine lettuce, along with vegetables like carrots and bell peppers, should be offered daily. These vegetables not only contribute to a balanced diet but also provide variety and enrichment for the rabbit's eating experience.

Commercial rabbit pellets are designed to supplement their nutritional needs, but they should be given in moderation. These pellets are formulated to ensure rabbits receive necessary nutrients such as fiber, protein, and vitamins. However, overfeeding pellets can lead to obesity, which can predispose rabbits to various health issues.

It's crucial to avoid feeding rabbits foods that are high in sugar or fat. These can disrupt their digestive system and contribute to obesity and other health problems. Fruits should also be given sparingly due to their high sugar content, even though rabbits may enjoy them as treats.

Hydration is equally vital for rabbits, especially during warmer months. Fresh, clean water should be available at all times to prevent dehydration and ensure proper organ function. Water bottles or bowls should be cleaned regularly to maintain hygiene and freshness.

Monitoring a rabbit's diet is essential for maintaining their health throughout their lifespan. Factors such as age, activity level, and overall health should influence feeding practices. Consulting with a veterinarian experienced in rabbit care can provide tailored advice based on individual nutritional needs and any specific health considerations.

Health And Disease Management

Maintaining the health and well-being of rabbits requires vigilant care and regular veterinary oversight to ensure they lead healthy lives. Rabbits are sensitive animals, and being attuned to signs of illness is crucial for early intervention and treatment. Key indicators of potential health issues include changes in appetite, behavior, or physical appearance, such as unusual discharge from the eyes or nose. Swift veterinary attention is essential upon noticing any of these symptoms to prevent complications and promote recovery.

Preventative healthcare plays a pivotal role in rabbit care. This includes adhering to a vaccination schedule designed to protect against common rabbit diseases like myxomatosis and viral hemorrhagic disease (VHD). Additionally, effective parasite control, typically through regular treatments for fleas, mites, and intestinal parasites, helps maintain their health and comfort.

Maintaining a clean living environment is another vital aspect of rabbit care. Regularly cleaning their habitat prevents the accumulation of bacteria and reduces the risk of infections, particularly in sensitive areas like their feet and hindquarters. Providing adequate space for exercise and mental stimulation is equally important. Rabbits are active animals that benefit from opportunities to hop, explore, and engage in natural behaviors, which helps prevent obesity and promotes physical and psychological well-being.

Nutrition is a cornerstone of rabbit health. A diet rich in hay, fresh vegetables, and controlled portions of pellets supports their digestive health and overall vitality. Monitoring their food intake and ensuring access to fresh water at all times helps prevent gastrointestinal issues and dehydration, common concerns in rabbits.

Regular veterinary check-ups are essential for comprehensive health monitoring. During these visits, veterinarians assess overall health, provide necessary vaccinations and parasite treatments, and offer guidance on diet and environmental enrichment tailored to the rabbit's specific needs. These proactive measures not only detect potential health issues early but also contribute to a longer and healthier life for pet rabbits.

Handling And Socializing Your Rabbits

Handling and socializing rabbits require gentle care and patience to ensure their well-being and foster trust between them and their owners. Proper handling techniques are crucial, primarily focusing on supporting their hindquarters to avoid spinal injury, as their spines are delicate. This approach not only prevents harm but also builds a foundation of trust.

Daily socialization is key to helping rabbits feel comfortable with human interaction and reducing fearfulness. Spending consistent, quality time with them through gentle petting, talking softly, and offering treats establishes positive associations with their human caregivers. This interaction should be approached calmly to avoid startling the rabbits, allowing them to gradually acclimate to human presence and touch.

Introducing rabbits to new environments and experiences should be done gradually to minimize stress and help them feel secure. Providing a safe, enriching environment within their living space is essential. This includes offering toys like tunnels and platforms that encourage physical activity and mental stimulation. These enrichments not only keep rabbits physically healthy but also prevent boredom, which can lead to behavioral issues.

Bonding with rabbits goes beyond physical care; it involves creating a nurturing and trusting relationship. This can be achieved through consistent routines, such as feeding and playtime, where rabbits learn to anticipate positive interactions with their owners. Understanding their individual personalities and preferences also plays a significant role in strengthening this bond.

Additionally, observing rabbits for signs of discomfort or distress is important. These signs may include thumping hind legs, freezing, or attempting to escape. Respect their cues and give them space when needed to build confidence in their interactions.

CHAPTER FIVE

BREEDING MEAT RABBITS

Understanding Rabbit Reproduction

Rabbits are renowned for their prolific breeding habits, making them highly suitable for meat production due to their rapid reproductive cycles. Understanding the intricacies of their reproduction is essential for successful breeding management.

Female rabbits, referred to as does, typically reach sexual maturity between 5 to 6 months of age. At this stage, they enter into a regular estrous cycle, also known as a reproductive cycle. This cycle generally recurs every 16 to 18 days, during which the doe becomes receptive to mating. This period of receptivity lasts for about 12 to 14 hours, making timing crucial for successful breeding.

Male rabbits, known as bucks, attain sexual maturity slightly earlier than does, typically becoming capable of mating at around 3 to 4 months old. Bucks are generally more sexually active and can service multiple does in a short period, contributing to their reputation as efficient breeders.

When planning to breed rabbits, ensuring the health and genetic fitness of both bucks and does is paramount. Genetic defects in either parent can be inherited by offspring, potentially leading to health issues or other complications in future generations.

Therefore, responsible breeding practices involve selecting rabbits with robust genetic backgrounds and ensuring they are free from hereditary health concerns.

Breeding rabbits can be managed through various methods, including natural mating or artificial insemination, depending on the specific breeding goals and circumstances. Natural mating involves placing a receptive doe with a sexually active buck during her estrous period. Careful observation of mating behaviors and ensuring a stress-free environment can improve breeding success rates.

Alternatively, artificial insemination offers advantages in controlled breeding programs, allowing breeders to select specific genetic traits and overcome logistical challenges such as distance or health considerations of the rabbits involved. This method requires expertise and careful handling to ensure the viability of the collected semen and successful insemination.

After successful mating, a doe's gestation period lasts approximately 31 days, during which she prepares for the birth of her litter. Proper nutrition and housing are crucial during this period to support the doe's health and ensure the development of healthy offspring.

Selecting Breeding Stock

Selecting the appropriate breeding stock is a critical step in ensuring the health and vigor of future generations of rabbits, especially when focusing on producing robust kits for meat production. The process involves careful consideration of several key factors that contribute to the overall quality and health of the offspring.

First and foremost, selecting rabbits with desirable genetic traits is essential. This includes evaluating factors such as growth rates, body conformation, and meat qualities. Rabbits with good growth rates tend to produce offspring that grow quickly and efficiently, which is beneficial in meat production operations where efficiency is paramount. Body conformation, including aspects like skeletal structure and muscle development, directly influences the meat yield and overall health of the rabbits.

Avoiding closely related rabbits in breeding is crucial to minimize the risk of genetic abnormalities and inherited diseases. Inbreeding can amplify genetic defects and reduce the overall vitality of the offspring. Therefore, breeders should carefully track the lineage of their rabbits and select breeding pairs that are genetically diverse to maintain healthy genetic variability.

Furthermore, ensuring the health of breeding rabbits is fundamental. Before selecting them for breeding, it's essential to conduct thorough health checks. Look for signs of disease or illness, such as abnormal discharge, lethargy, or respiratory issues. Healthy rabbits are more likely to produce healthy offspring, so maintaining rigorous health standards is key to sustainable breeding practices.

In addition to genetic and health considerations, the diet and living conditions of breeding rabbits play a significant role in their reproductive success. A balanced diet rich in nutrients supports optimal reproductive health and ensures that rabbits are in peak condition for breeding. Adequate living conditions, including sufficient space, proper ventilation, and cleanliness, contribute to the overall well-being and stress levels of rabbits, which can directly impact their reproductive performance.

Successful breeding operations also involve continuous monitoring and evaluation. Regularly assess the performance of breeding stock and their offspring to identify any areas for improvement or adjustment in breeding strategies. This ongoing evaluation helps maintain and improve the quality of the breeding stock over time.

Mating And Gestation

Mating and gestation are critical phases in the reproductive cycle of rabbits, requiring careful observation and preparation to ensure a successful outcome. When the doe enters estrus, commonly known as being "in heat," she displays distinctive behaviors indicating her readiness to mate. These behaviors include increased activity and a willingness to be mounted by the buck. Given that females can be territorial, it's advisable to introduce the buck into the doe's cage or territory rather than relocating the doe, which could cause stress.

Supervising the mating process is essential to prevent aggression between the rabbits and to ensure that successful copulation occurs. This supervision helps maintain a safe environment for both animals, minimizing the risk of injury or stress-related complications. It also allows for the monitoring of mating behavior, ensuring that the doe is properly impregnated.

Following successful mating, the gestation period in rabbits lasts approximately 31 days. During this time, it's crucial to provide the pregnant doe with a stress-free environment. Stress can negatively impact the pregnancy, potentially leading to complications or even miscarriage. Maintaining a calm atmosphere and minimizing disturbances are key to supporting the doe's health and the development of her offspring.

As the due date approaches, preparing a suitable nesting area becomes imperative. A clean, comfortable nesting box should be provided, ideally filled with soft hay or straw. This material serves not only to provide warmth and comfort but also to encourage natural nesting behaviors in the doe. Ensuring the nesting box is adequately sized and placed in a quiet area further supports the doe in preparing for birth.

Monitoring the pregnant doe's health throughout gestation is essential. Regularly checking her food and water intake, as well as observing any changes in behavior or physical condition, allows for early detection of potential issues. Prompt veterinary attention should be sought if any concerns arise, ensuring the doe receives appropriate care.

☐

Caring For Pregnant Does And Newborn Kits

Caring for pregnant does and their newborn kits is crucial to ensuring their health and well-being. Proper management includes monitoring the doe's condition, preparing for birth, and providing postnatal care for the kits.

During pregnancy, it's essential to monitor the doe's weight and adjust her diet accordingly. Increasing her protein intake helps support the developing kits and prepares her body for milk production. As the pregnancy progresses, typically around the 28th day of gestation, the doe will exhibit nesting behavior by pulling hair from her chest and abdomen to create a warm nest for her impending litter. It's important to provide a nesting area that is warm and draft-free to accommodate the delicate newborns, who are highly sensitive to temperature fluctuations.

Newborn kits are born hairless and blind, relying entirely on their mother for warmth, nutrition, and care. The doe will nurse them once or twice a day, usually during the quieter times of dawn and dusk to minimize attracting predators. This nursing schedule is crucial for the kits' growth and development.

Kits should remain with their mother for at least 6-8 weeks to ensure they are fully weaned and capable of consuming solid food independently. During this period, regularly check the kits for any signs of illness or malnutrition. Handling them gently during these checks minimizes stress on both the kits and the mother, promoting a healthy environment for their growth.

Maintaining a clean and hygienic environment is also essential. Ensure the nesting area remains clean to prevent infections and keep the doe comfortable. Providing fresh water and a balanced diet for the mother helps ensure she has the necessary nutrients for milk production and overall health.

Observing the doe's behavior and monitoring the kits' progress is key to early detection of any health issues. If you notice any abnormalities, such as weight loss in the kits or signs of distress in the mother, consult a veterinarian experienced in caring for rabbits promptly.

CHAPTER SIX

GROWING AND HARVESTING

Growth Stages Of Meat Rabbits

Meat rabbits progress through several distinct growth stages, each crucial for their development and eventual readiness for harvesting. Understanding these stages is essential for efficiently managing their growth and ensuring optimal meat production.

Kit Stage: The journey of a meat rabbit begins at birth, where newborns are known as kits. At this stage, typically lasting from birth to around 4 weeks old, kits are entirely reliant on their mother's milk for nutrition and warmth. This period is critical for their initial growth and development, as they gain strength and begin to explore their environment under the watchful care of the mother rabbit.

Weaning Stage: Around 4 to 6 weeks of age, kits start to transition from solely nursing to nibbling on solid foods. This gradual introduction to solid food marks the weaning stage. It's a pivotal time when the kits learn to consume a more varied diet, supplemented by the nutrients they continue to receive from their mother's milk. Care during this period ensures that the transition to independent feeding goes smoothly, supporting healthy growth and development.

Grow-out Stage: From approximately 6 weeks to 12 weeks of age, meat rabbits enter the grow-out stage. During this phase, they experience rapid growth, putting on weight and developing muscle mass. Proper nutrition becomes increasingly important during this period to support their growth rate and overall health. The goal is to ensure they reach a size and weight that are optimal for meat production, setting the foundation for the next stage of their development.

Finishing Stage: By 12 to 14 weeks of age, rabbits are approaching their mature size and weight in the finishing stage. This stage focuses on maximizing their growth potential and ensuring they are ready for harvest. Attention to nutrition, environmental conditions, and health monitoring becomes critical during this final phase before processing. The aim is to achieve consistent growth and development, producing rabbits that meet the desired standards for meat quality and quantity. Each of these growth stages plays a vital role in the lifecycle of a meat rabbit, from the initial dependency on maternal care to the independent feeding and rapid growth phases leading up to harvest. Effective management practices, including nutrition management, hygiene, and health monitoring, are essential throughout these stages to optimize growth efficiency and ensure the production of high-quality meat rabbits.

Understanding and managing these growth stages enable rabbit breeders to maximize productivity while maintaining the health and welfare of their animals.

Optimal Conditions For Growth

Creating optimal conditions for the growth and health of rabbits is crucial for maximizing their well-being and productivity. Several key factors play a significant role in achieving this goal, ranging from housing and nutrition to hydration and healthcare.

Firstly, housing plays a pivotal role in providing rabbits with a conducive environment. It is essential to offer spacious hutches or cages that are well-ventilated. This not only ensures adequate air circulation but also protects the rabbits from extremes of weather and potential predators. The housing should be designed to provide both shelter and security, allowing the rabbits enough space to move comfortably without overcrowding, which can lead to stress and health issues.

Nutrition is another critical aspect of rabbit care. A balanced diet is essential for their growth and overall health. The diet should be rich in fiber, which supports digestive health, along with sufficient protein and essential nutrients. This typically includes a combination of high-quality hay, specially formulated pellets, and fresh greens. Hay, in particular, is crucial as it helps maintain proper dental health and digestive function in rabbits. Pellets provide concentrated nutrition, while fresh greens add variety and additional nutrients to their diet.

Hydration is equally important. Rabbits should always have access to fresh, clean water. Water is essential for digestion, nutrient absorption, and overall hydration, which supports their metabolic processes and helps regulate body temperature. Providing a constant supply of water ensures that rabbits remain healthy and prevents dehydration, especially during warmer months or periods of increased activity.

Healthcare is another aspect that should not be overlooked. Regular monitoring for signs of illness, injury, or parasites is essential. Early detection allows for prompt intervention and treatment, minimizing potential health issues and ensuring the rabbits' well-being. Veterinary care should be sought as needed, including regular check-ups and vaccinations to prevent common diseases. This proactive approach to healthcare helps maintain the rabbits' overall health and longevity.

Determining The Right Time For Harvest

Determining the optimal time for harvesting rabbits involves careful consideration of several key factors to ensure both quality and efficiency in meat production. The decision revolves around age, weight, body condition, and market demand.

Firstly, age and weight play crucial roles in determining the readiness of rabbits for harvest. Generally, rabbits are harvested between 10 to 14 weeks of age, depending on breed and feeding regimen. This period allows them to reach their peak weight for meat production. Young rabbits are prized for their tender meat and optimal flavor, making this age range ideal for maximizing meat yield per animal.

Secondly, assessing the body condition of rabbits is essential before harvest. A well-developed rabbit should exhibit good muscle mass and overall health. Visual inspection and palpation can help determine if the rabbits have reached sufficient maturity and muscle development for slaughter. Rabbits at this stage typically have a well-rounded body shape and firm muscle tone, indicating readiness for processing.

Market demand also significantly influences the timing of rabbit harvest. Understanding market preferences and seasonal demand patterns allows producers to time their harvests effectively. By aligning production cycles with peak demand periods, such as holidays or special events, farmers can optimize sales and profitability. This strategic approach ensures that rabbits are harvested when there is a robust market appetite, minimizing storage time and ensuring freshness for consumers.

Moreover, environmental factors may influence the decision on when to harvest rabbits. Seasonal variations in temperature and feed availability can impact growth rates and overall readiness for harvest. Monitoring these factors helps farmers adjust their production schedules to maximize efficiency and minimize costs.

Humane Methods Of Harvesting
Humane harvesting practices are crucial for ensuring the welfare and respectful treatment of rabbits throughout the harvesting process. Central to this

approach is the use of approved stunning methods, such as cervical dislocation or stunning with a captive bolt, to effectively render rabbits unconscious before processing begins. This step is essential as it minimizes any potential pain or distress the rabbits might experience during handling and slaughter.

Once stunned, it is imperative to proceed with quick and efficient processing. This not only ensures minimal suffering but also maintains the quality of the meat. Swift processing reduces the time rabbits spend in stressful conditions, thereby contributing to a more humane treatment overall.

Creating an appropriate environment is another key aspect of humane harvesting. Maintaining a calm and quiet atmosphere during handling and processing helps to reduce stress levels in the rabbits. Stress can negatively impact meat quality and is also ethically undesirable, as it affects the well-being of the animals. By minimizing stress through a peaceful environment, handlers can uphold standards of humane treatment.

Respectful handling is fundamental throughout every stage of the harvesting process. Handlers should approach rabbits gently and with care, acknowledging their responsibility for the animals' welfare. This approach not only aligns with ethical considerations but also contributes to the overall quality of the end product. Treating animals with respect is not only morally right but also ensures that the meat produced is of high quality and free from avoidable blemishes or defects caused by undue stress.

Furthermore, adherence to humane harvesting practices is not only a matter of ethical responsibility but also reflects positively on the entire agricultural and food production industry. Consumers increasingly prioritize products sourced from animals treated with care and respect. Therefore, implementing and promoting humane practices can enhance the reputation of producers and contribute to consumer confidence in the products they offer.

CHAPTER SEVEN
PROCESSING RABBIT MEAT
Preparing For Processing

Preparing for processing meat is a crucial step to ensure food safety and efficiency in the kitchen. Before you start, it's essential to gather all the necessary tools and prepare your workspace thoroughly. Here's a detailed guide to help you get organized:

Firstly, gather your tools. You'll need a sharp knife, preferably one suited for cutting meat to ensure clean cuts and reduce shredding. A cutting board designated specifically for meat is essential to prevent cross-contamination with other foods. Containers with lids for storing meat during and after processing are necessary to maintain freshness and prevent exposure to bacteria.

Secondly, ensure your workspace is clean and sanitized. All surfaces, including countertops, cutting boards, and utensils, must be thoroughly cleaned with hot, soapy water and sanitized with a solution of bleach and water or a kitchen sanitizer. This step is crucial to eliminate any bacteria that could potentially contaminate the meat.

Next, organize your workspace for efficiency. Arrange your tools and containers within easy reach to streamline the processing process. This organization not only saves time but also reduces the risk of accidents or contamination from reaching across surfaces unnecessarily.

Before handling meat, wash your hands thoroughly with soap and warm water for at least 20 seconds. Proper handwashing is critical to prevent the spread of bacteria from your hands to the meat.

Once everything is set up, proceed with preparing the meat. Start by inspecting the meat for any signs of spoilage, discoloration, or unusual odors. Trim off any visible fat or connective tissue as needed. Use the sharp knife to make precise cuts according to your recipe or meal plan.

As you work, maintain a clean workspace by immediately disposing of any scraps or trimmings into a designated trash container. This practice helps prevent cross-contamination and keeps your work area tidy.

Throughout the processing, periodically sanitize your tools and surfaces to ensure continued cleanliness. This practice is especially important if you handle different types of meat or switch between raw and cooked foods.

Finally, after processing is complete, store the meat in clean, airtight containers in the refrigerator or freezer promptly to maintain freshness and prevent bacterial growth. Label containers with the date to keep track of storage times.

Step-By-Step Processing Guide

Raising and processing rabbits for meat requires careful attention to humane practices and precise handling to ensure food safety and quality. Follow these steps to process rabbits efficiently and effectively:

Humane Slaughtering: The process begins with humane slaughtering methods, such as cervical dislocation or stunning followed by bleeding out. These methods are chosen to minimize stress and pain for the rabbits, aligning with ethical standards and regulatory guidelines.

Skinning: After humane slaughter, the rabbit is prepared for skinning. Begin by making a small incision around the ankles, taking care to avoid damaging the underlying flesh. Slowly peel back the skin, working methodically to prevent tearing. This process requires patience and skill to ensure the hide is removed cleanly without compromising the meat.

Evisceration: Once skinned, proceed to evisceration. Make a precise incision from the pelvis up to the rib cage, ensuring the cut is clean to avoid puncturing internal organs. Carefully remove the entrails, taking caution not to rupture the intestines or other organs. This step is critical for maintaining the cleanliness and quality of the meat.

Rinsing: With the entrails removed, thoroughly rinse the carcass with cold water. This step helps to cleanse the meat of residual blood and debris, promoting hygiene and preparing the rabbit for further processing.

Final Inspection: Conduct a meticulous final inspection of the carcass. Check for any remaining hair or contaminants that may have been missed during previous steps. Trim any excess fat or connective tissue to ensure the meat is clean and ready for consumption.

Packaging and Storage: After inspection and trimming, rabbits can be portioned and packaged according to your preference. Use clean, food-grade packaging materials to maintain freshness and prevent contamination. Proper labeling with date and type of meat ensures clear identification during storage and distribution.

Processing rabbits for meat production demands attention to detail, adherence to humane practices, and strict hygiene protocols. By following these steps—from humane slaughtering to final inspection—you can ensure the meat is safe, clean, and of high quality for consumers. Each stage requires skill and precision to handle the rabbits respectfully and efficiently, reflecting both ethical considerations and practical aspects of meat processing.

Meat Cuts And Preparation

When preparing rabbit meat, understanding various cuts and preparation methods ensures optimal flavor and tenderness in your dishes. Here's a comprehensive guide to handling and preparing rabbit meat:

Whole Rabbit: One approach to preparing rabbit is leaving it whole, which is ideal for roasting or grilling. Keeping the rabbit intact helps preserve its natural flavors and ensures even cooking. This method is straightforward and highlights the simplicity and delicacy of rabbit meat.

Quartering: For more controlled cooking and serving, quartering the rabbit is a practical method. This involves dividing the rabbit into four main parts: the front legs, hind legs, and the saddle (the back portion). Quartering allows for even cooking of each section and facilitates different cooking techniques for each part, such as braising or grilling.

Cutting into Pieces: To expand the culinary possibilities, cutting the rabbit into smaller pieces is advantageous. This can include separating the rabbit into smaller portions suitable for stews, frying, or other cooking methods. Ensuring uniformity in the size of the pieces helps achieve consistent cooking times and ensures that each piece cooks evenly.

Marinating (Optional): Marinating rabbit pieces is optional but can significantly enhance flavor and tenderization. A marinade typically consists of a mixture of herbs, spices, and oils. Marinating the rabbit pieces for several hours to overnight allows the flavors to penetrate the meat, resulting in a more flavorful dish. Refrigeration during marination helps maintain food safety and ensures the rabbit absorbs the marinade effectively.

Each method of preparing rabbit meat offers distinct advantages depending on your culinary preferences and the desired dish. Whole rabbit preparation is straightforward and preserves the natural integrity of the meat, ideal for those who appreciate simplicity and natural flavors. Quartering facilitates more controlled cooking and presentation, suitable for dishes where different parts of the rabbit can be showcased individually. Cutting into pieces enhances versatility, allowing for various cooking techniques and culinary creations.

Marinating rabbit meat adds another layer of flavor and tenderness, making it a worthwhile step for those looking to elevate their rabbit dishes. Whether you choose to roast a whole rabbit, quarter it for grilling, cut it into pieces for stews, or marinate it for enhanced flavor, understanding these methods ensures a delicious outcome. Experimenting with different preparations can help you discover new ways to enjoy this delicate and nutritious meat, making rabbit a versatile choice for both everyday meals and special occasions.

Storing And Preserving Rabbit Meat

Storing and preserving rabbit meat is essential to maintain its quality and safety for consumption over time. Here's a comprehensive guide on different methods you can use:

Refrigeration: Fresh rabbit meat should be stored promptly after purchase or slaughter to maintain freshness. Place it in a sealed container or vacuum-sealed bag to prevent exposure to air and potential contamination. In the refrigerator, rabbit meat can be safely stored for 2-3 days. Ensure the refrigerator is set to 40°F (4°C) or below to inhibit bacterial growth and maintain quality.

Freezing: For longer-term storage, freezing rabbit meat is highly effective. Start by wrapping the meat tightly in plastic wrap to minimize exposure to air. This step prevents freezer burn and helps retain moisture. Then, wrap the meat in foil for added protection or use vacuum-sealed bags to remove as much air as possible. Label each package with the date of freezing to track freshness. Frozen rabbit meat can be stored for up to 6 months without significant loss of quality if kept at 0°F (-18°C) or below.

Canning (Optional): An alternative method for preserving rabbit meat is canning, which allows for long-term storage at room temperature. Use proper canning techniques to ensure safety and longevity. Start by cutting the rabbit meat into suitable pieces and packing them into sterilized jars. Follow a reliable canning recipe and processing method to eliminate bacteria and create a vacuum seal. Properly canned rabbit meat can be stored for an extended period, typically up to one year or more, depending on the specific canning process used.

Cooked Rabbit: Once rabbit meat is cooked, it should be stored with equal care to maintain its flavor and safety. Store cooked rabbit in the refrigerator in airtight containers for 3-4 days. If you plan to store it longer, freezing is again an excellent option. Place cooled cooked rabbit in airtight containers or freezer bags, ensuring they are properly sealed to prevent freezer burn. Frozen cooked rabbit maintains quality for up to 3 months.

CHAPTER EIGHT

SUSTAINABLE AND ETHICAL PRACTICES

Ethical Considerations In Rabbit Farming

Ethical considerations in rabbit farming are paramount to ensuring the welfare and humane treatment of rabbits throughout their lives. Central to these considerations are practices that prioritize the rabbits' physical and psychological well-being, aiming to minimize stress and discomfort at every stage of their care.

Firstly, proper housing is fundamental. Rabbits require adequate space to move freely and exhibit natural behaviors. Cages should be spacious enough to accommodate their size, allowing room for movement and exploration. Clean bedding is essential for hygiene, and shelters provide protection from extreme weather conditions, ensuring rabbits remain comfortable and safe.

Nutrition plays a critical role in rabbit health. A balanced diet is crucial, typically consisting of quality commercial pellets that meet their nutritional needs. Supplementing with fresh greens and hay supports their digestive health and provides essential fiber. Access to clean water at all times is also essential to prevent dehydration and maintain overall health.

Healthcare is another key aspect of ethical rabbit farming. Regular veterinary check-ups are necessary to monitor their health and catch any potential issues early. Vaccinations and parasite control measures are implemented to prevent diseases, ensuring rabbits live healthy lives free from unnecessary suffering.

Breeding practices in ethical rabbit farming focus on responsible management to avoid overbreeding and genetic problems. Farmers prioritize the health and genetic diversity of both parent rabbits and their offspring. Controlled breeding programs ensure that only healthy rabbits with desirable traits are selected for reproduction, maintaining the overall quality and health of the herd.

Handling and transport procedures are carefully managed to minimize stress. Gentle handling techniques are employed during routine care and transportation to reduce anxiety and potential injury. Proper crate design and ventilation in transport vehicles ensure rabbits are transported safely and comfortably, minimizing the risks associated with travel.

Overall, ethical rabbit farming requires a commitment to providing compassionate care and ensuring the highest standards of welfare. By adhering to these principles, farmers not only promote the health and happiness of their rabbits but also contribute to sustainable and responsible agricultural practices. Ethical considerations are integral to fostering a positive relationship between humans and animals, acknowledging the inherent value of every rabbit under their care.

Implementing Sustainable Practices

Sustainable rabbit farming focuses on minimizing environmental impact and maximizing resource efficiency through a variety of practices. These practices are designed to not only reduce operational costs but also contribute positively to the environment.

Efficient resource use lies at the core of sustainable rabbit farming. By optimizing water, feed, and energy consumption, farmers can minimize waste and lower their overall operational costs. This approach not only improves economic sustainability but also reduces the ecological footprint of rabbit farming operations.

Waste management plays a crucial role in sustainable practices. Recycling and composting rabbit waste are effective strategies to minimize environmental pollution. Rabbit manure, for instance, can be composted and used as organic fertilizer, thereby closing the nutrient loop and reducing the need for synthetic fertilizers.

Adopting alternative energy sources is another hallmark of sustainable rabbit farming. By harnessing renewable energies such as solar or wind power, farmers can reduce their reliance on fossil fuels. This shift not only decreases greenhouse gas emissions but also enhances the long-term viability of farm operations by stabilizing energy costs.

Biodiversity conservation is integral to sustainable farming practices. By preserving natural habitats within and around farm areas, farmers can support local ecosystems and promote biodiversity. This conservation effort contributes to soil health, enhances pollination services, and supports natural pest control mechanisms.

Integrated pest management (IPM) techniques are preferred over conventional chemical methods in sustainable rabbit farming. IPM utilizes natural predators, beneficial insects, and companion planting to control pests effectively while minimizing environmental impact. This approach not only preserves ecosystem balance but also reduces the risk of pesticide residues in rabbit products.

Overall, sustainable rabbit farming integrates these practices to achieve a balanced approach to agriculture. By prioritizing environmental stewardship and resource efficiency, farmers can ensure the long-term viability of their operations while contributing positively to broader sustainability goals. These practices not only benefit the environment but also enhance farm resilience to future challenges such as climate change and resource scarcity.

Utilizing Rabbit By-Products

Utilizing rabbit by-products efficiently is essential for promoting sustainability and economic viability across various industries. Rabbit meat, renowned for its leanness and nutritional benefits, meets a growing demand in the market. Ethical processing practices prioritize humane slaughter methods, ensuring both the quality of the meat and the ethical treatment of animals. This approach not only satisfies consumer preferences for high-quality products but also aligns with ethical standards in food production.

In addition to meat, rabbit fur plays a significant role in the fashion industry. Ethical fur harvesting practices ensure minimal waste and adhere to humane standards, contributing to the sustainable use of resources. Rabbit fur is utilized in clothing and accessories, valued for its softness and insulation properties. By integrating ethical practices, the industry supports both environmental conservation and consumer demand for sustainable fashion choices.

Rabbit manure emerges as a valuable resource in agriculture. Rich in nitrogen and phosphorus, it serves as an organic fertilizer that enhances soil fertility naturally. Unlike chemical fertilizers, rabbit manure contributes to sustainable farming practices by reducing reliance on synthetic additives and promoting soil health over the long term. This eco-friendly approach supports sustainable agriculture and mitigates environmental impact through natural soil enrichment methods.

Furthermore, rabbit tissues and organs are vital in biomedical research and pharmaceutical development. Their biological similarity to humans makes rabbits invaluable in studying diseases, testing treatments, and developing pharmaceutical products. Ethical considerations in research ensure the humane treatment of animals while advancing medical science, highlighting the dual benefits of ethical practices in both research ethics and scientific progress.

Engaging With The Community And Educating Others

Engaging with the community and educating others about ethical and sustainable practices in rabbit farming is crucial for fostering understanding and garnering support. This multifaceted approach involves education programs, local market involvement, and collaboration with various stakeholders.

Education programs play a pivotal role in informing the public about rabbit farming practices and their broader benefits. Through workshops, farm tours, and educational materials, we aim to demystify the complexities of ethical animal husbandry. Workshops provide hands-on learning experiences, where participants gain insights into the daily operations of rabbit farming, from housing and feeding to healthcare and breeding practices. These sessions not only educate but also inspire a deeper appreciation for sustainable agriculture.

Farm tours complement workshops by offering firsthand experiences on-site. Visitors witness sustainable farming practices in action, such as composting, rotational grazing, and natural pest management. These tours highlight our commitment to transparency and ethical standards, allowing the community to see the care and dedication that goes into every aspect of rabbit welfare.

Local markets serve as more than just venues for selling products; they are hubs for community interaction and support. By selling locally produced rabbit meat and other products, we not only provide fresh, high-quality food but also strengthen community ties. Engaging with consumers directly allows us to share our values and educate them about the advantages of choosing locally sourced, ethically raised products. This direct connection fosters trust and encourages sustainable consumption practices among community members.

Collaboration is another cornerstone of our community engagement strategy. Partnering with schools, environmental organizations, and government agencies allows us to amplify our impact and reach broader audiences. By integrating rabbit farming education into school curricula, we nurture future generations' understanding of sustainable agriculture and animal welfare. Collaborating with environmental organizations helps us implement eco-friendly practices, ensuring that our farming methods minimize environmental impact. Working alongside government agencies supports regulatory compliance and promotes policies that prioritize ethical animal husbandry.

THE END

www.ingramcontent.com/pod-product-compliance
Lightning Source LLC
Chambersburg PA
CBHW071833210526
45479CB00001B/112